For

Dad who instilled a love of Science

and

Mum who always encouraged my Art.

Jenna Whyte
2 Cambridge Drive
Leeds

The Illustrated Guide to the Elements

First published in
Great Britain 2012
By
Jenna White

Any enquires concerning reproduction of this book should be emailed to
Jennawhyte@hotmail.com

ISBN
978-0-9575327-0-0

www.Jennawhyte.co.uk

Second Edition
20/02/13

Cover illustration by Jenna Whyte
Cover design by Holly Trafankowska

The Illustrated Guide to the Elements

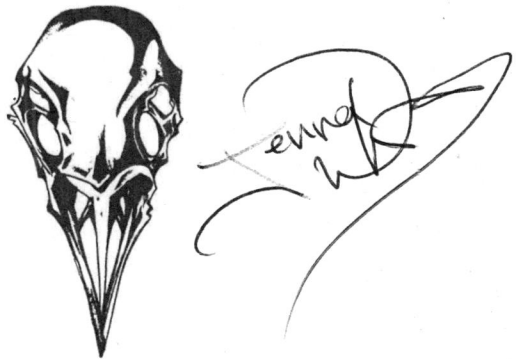

by Jenna Whyte

PRUSSIAN BLUE INK

The perfect antidote for heavy metal poisoning
Stopping Thallium and Radio Active Caesium in their tracks!

PROTECT YOUR SELF!

Every family should have a bottle in their cupboard

! Beware of Spurical and Piratical Imitations!

POST FREE
PRICE 2'9 PER BOTTLE

We do not take responsibility for any deaths incurred ⋎ this product

Elements Volume I
Contents

To whom it may concern:

This volume contains my findings on the women who make up the Elemental world.
I have focused my attention on the following Groups: The Alkaline Metals. the Halogens, the Post Transition Metals, the Metalloids and the Non-Metals.
The aim of this compendium is to show a glimpse inside a world driven by the
buying, selling, battling, begging, stealing and sharing of electrons, as well as the
personal properties of these women and any other notes that I have uncovered in my
investigation. Each file also includes a hand painted portrait of each femme fatale in
all their glory.

The deeper I delve into this investigation the more intrigued I become and the more
remarkable my findings. Elements that have always been considered devious and
poisonous have shown that they too can work for good and are in fact vital to human
life. Others once believed to be honest, spiritual and life affirming have turned out to
be quite the reverse. Some Elements go about their business quietly and unnoticed
by the hustle and bustle of the world, while some like to make a fanfare of their
endeavours, working their way into the lives and indeed hearts of the people - some
more literally than others.
Either way it is my wish that the data I have uncovered will lead you through this
world with caution where needed. respect when due, appreciation where appropriate
and a little more understanding of how these Ladies work, and how they effect our
lives.

Jenna Whyte

Jenna Whyte

ATTENTION!

! You that have a mind, be on the look out for the Insane !

Does someone near to you have sudden explosive fits of depression or mania? Are they volatile and unpredictable? Violent or overly introverted? Do they show an abnormal amount of attention in their electrons or electrons of others?
If so they could be insane.

CONTACT:
THE ASYLUM FOR ELECTRON CHALLENGED ELEMENTS

Be vigilant they could be lurking any where, any class
Whether it be the higher classes driven to madness by
the pressures of Upper-class life or the possession of a
womb which often drives the fairer sex to insanity.
All are at risk!

A PLACE FOR THE MAD TO BE SWEPT OUT OF SIGHT AND OUT OF MIND

The Alkaline Metals Group I

The Alkaline Metals are one of the more manic, doleful, and unstable Elemental groups. They are not well suited for the external world and appear to be able to cope with very little outside of their safe environment. For many of the Ladies in this category that 'safe environment' is "The Asylum for Electron Challenged Elements", which is the best place for them. Not just for our own safety as humans but also for theirs, as they have an unhealthy tendency of self combusting. These Elements must be handled with great care as they have very fragile temperaments.

One should never attempt to interact with them without the right precautions and equipment, or the result would be an attack of a feverish nature which could cause severe burns to the epidermis.

Research dictates that we should look deeper into the psyche of these morose Elements to ascertain the reasons why they act with such vigour when in water, air and almost every thing else. Perhaps then this knowledge may reveal why they are such sullen and miserable creatures.

It all comes down to their biology or electron figuration (see notes). They all have one electron in their outer shell, and they are all desperate to dispose of it. This makes them rather anti-social Elements, even without their erratic and disconsolate behaviour.

If we consider the periodic table as a map of the Elements and look down the Alkaline Metal group we see the Elements further down become even more aggressive as the electrons face less pull from their nucleus, thanks to their bigger mass and greater electron shielding. Additional characteristics that mark this family apart are, though they are metals they are all remarkably soft in body and soul. They can be easily cut with a knife; if a laceration is made they will shine for mere moments before the world tarnishes them and makes them dull. Other family traits include an inability to withstand heat. Once again for metals they have extremely low melting and boiling points, this decreases further still as you look down the group.

Further findings follow for individual Elements.

Lithium 3

Name: from Greek, 'lithos' meaning 'Stone'

Subject Notes:

Miss Lithium is a romantically tragic Element; much like Shakespeare's Ophelia she has little control over her body, relationships or choices due to her electron condition (as discussed in the chapter brief). Because of her Alkali Metal genealogy she is the most lethargic in her reaction compared to her sisters. She is the only one of her family who will resist reacting with Oxygen (unless heated). She gives up her outer electron easily with very little fight as she hides from the world behind her shells, afraid of the confrontation that may ensue. She has a perpetual air of melancholy and like many who suffer from mood disorders can act irrationally and explosively, particularly in water and is flammable.

Lithium is a sensitive, weepy soul and dislikes being out in public. As a result of such exposure she quickly loses her silvery lustre in air, becoming dull and tarnished, eventually turning black. She is also the lightest solid Element, adding to her delicate composure.

Subject 3 has moments when she is escorted to the 'The Asylum for Electron challenged Elements' but is soon released, and for the most part manages to stay out, albeit under the watchful eyes of the psychiatric attendants who watch as she aimlessly broods around her family's estates.

Subject History:

Miss Lithium is a one of the oldest Elements, being among the first to emerge from the explosion of the big bang. Subject 3 is used in mood stabilizers for the bi-polar. Unlike the other mood altering drugs, she can be used to treat both depression and mania. She is, however, a far cry from the clichéd 'happy pill'; being most effective at treating mania in bi-polar patients.

When first tested on guinea pigs the creatures slowed right down and became completely docile. After this, dangerously high doses were given to the most disturbed and manic patients who had been institutionalised for years. The result was that over a relatively short period of time with Lady Lithium, they were ready to move to a regular hospital and not long after released back into civilised society.

Lithium is moderately toxic; in high doses she can cause confusion and slurred speech and eventually death. Some humans have a higher tolerance for her than others.

She is also corrosive and will burn skin as she reacts with the sweat hands, evidence that this wistful Element abhors to be touched and shuns physical contact as well as company.

She does have other none biological roles in which her light weight comes into play. Lithium can be used in alloys to lighten them, being useful in aircraft.

3

Li

Lithium

Sodium 11

Name: From the English word 'soda' which in turn came from the Medieval Latin word 'sodanum', meaning "headache remedy." Sodium's chemical symbol NA comes from the Latin word for Sodium carbonate, 'natrium'

Subject Notes:

Sodium is probably one of the most notorious, used, and debated Elements. Like all Alkaline Metals she is soft, easily tarnished, light weight and very reactive.

She is essential for life, without her our bodies would not be able to regulate osmotic pressure and organic acids with the blood. Poor, unfortunate Sodium is too easily flushed out of the body; this can be problematic as the body cannot make its own supply of the vital salts Sodium supplies. Those salts must be replaced, keeping Sodium busy at all times. One of her most important jobs is to keep the movement of electrical impulses constant, and she does this in tandem with Miss Potassium. Potassium is most favoured of all her sisters, the two sisters are very close and have endured a well-to-do childhood together and often shared cells together in 'The Asylum for Electron Challenged Elements'. Despite Sodium's usefulness it is always in debate however as to how much of this Element should be consumed. Many argue that excessive salt will cause illness, mostly in the form of high blood pressure. Others disagree with this hypothesis stating that it only exacerbates high blood pressure. They also maintain that the more salt one has the longer they will live, using the Japanese who have the longest life span as an example. It is no wonder then that this unstable and nervous Element suffers bouts of confusion, hysteria and paranoia, never knowing where she should be. Another way in which Sodium influences our lives daily is through her most extraordinary property that has gained her a universal reputation: that of taste. Sodium-Chloride is the only mineral that triggers its own reaction on our tongues, one of our basic tastes – salt.

When ignited or subjected to an electric current (a method to 'cure' the mentally ill) she will burn a bright outlandish orange hue. It is thanks to this property we, especially in the western world, get our amber street lights. Though Patient 11 is providing a public service lighting our roadways and aiding our journeys, many dislike the lights, complaining of light pollution: An orange eyesore that can be seen for miles, obstructing celestial lights of the night sky. No matter how much Sodium does for us, there is always some drawback and those who will question her motives and work, pushing her further and further into a dark depression and causing her to question her self at every step. It is my conclusion that one should take care when encountering her not only because of her explosive nature, but also because of her more sensitive feelings. It is these attributes that insure she will remain incarcerated at the asylum to the end of her days; only allowed escape from her confines to exercise her skills and work her fingers to the bone.

11 NA

Sodium

Potassium 19

Name: From the English word 'potash', the symbol K is from the Latin word 'kalium' meaning 'alkali'

Subject Notes:

Potassium is a fragile Element at best, always on the verge of a reaction or break-down in order to free herself of an electron. Despite her questionable psychosis, she is part of the Sodium-Potassium powerhouse who work as a team to keep the human body functioning. This, nevertheless, takes its toll on this sombre Element leaving her overworked, neurotic and as hysterical as her sisters, the pressure of public demand weighing heavily on her shoulders. She is unfortunately another patient in the 'Asylum for Electron challenged Elements', and is only permitted out of the premises to complete her tasks and go about her work. She has not exhibited any behaviour that suggest her release will be imminent and thus for the foreseeable future will spend the rest of her days as a resident in the asylum.

Subject History:

This Element's salts were known before the official Element was discovered by Sir Humphrey Davy in 1807. In the form of ash, Potassium was used by Native Americans to flavour and preserve food; it was also used in 17th Century Asia to improve soil quality. To this day 95% of Potassium is used in fertilizers; a role favoured by none of the Elemental world, a fact no doubt depresses Patient 19 further.

She is also utilized in many other areas of our lives. For example: when added to glass it becomes scratch resistant, she is used in baking, tanning leather, iodizing salt and photography.

Most importantly Potassium is essential to almost every living thing. In humans she is vital to red blood cells, muscle functions, keeping the kidneys in working order, making tissue, regulating fluids and operating nerve impulses.

In spite of all this there is a darker side to this Element. Murders have been committed with her assistance, and overdose of this vital Element will cause a depression on the central nervous system, cause convulsions, diarrhoea, kidney failure and eventually death. Wrongly or rightly doctors and nurses have ended lives of sick patients with a concoction of Potassium and a Chloride. This is also the method used in capital punishment, the lethal injection for those whose organs will be harvested after death for transplants.

19 K

K·19

Potassium

Rubidium 37

Name: from Latin 'rubidius' meaning Deepest Red

Subject Notes:

This tortured, listless creature cannot function in society or in any environment out-side of that provided by the asylum. She will react with almost everything.

One would not attempt to awaken this Element's senses with the age old method of repetitive submersion in water for her reaction would be less than favourable, as this hapless Element will react vigorously with water and ice.

Patient 37 must remain locked up at all times and will never be released from of the Asylum, for any reason. As stated in the beginning of this chapter this precaution is for the safety of all involved. Oil or grease works best to contain her as in open air she may combust at any given moment. She may not be brought into contact with any other Element for the same reasons. She seems set to spend the rest of her days in solitude and lonely confinement.

Subject History:

Miss Rubidium is easily absorbed by our human bodies, due to the fact she is much like her sister Potassium by nature. Once there she may have a slight stimulatory effect on the metabolism because of this similarity. Other than this she has no effect on us, she is neither nutritious nor toxic.

She is a rare Element and is, as a result very expensive, her favours costing more than Platinum or Gold per kilogram.

She is often out of work, as there is little use for her, and what use there is, is re-stricted to research in laboratories.

What makes her situation all the more heartbreaking is that nobody wants to find a use for this dejected Element. Any job that Miss Rubidium is suited for can easily be done by Sodium or Potassium at half the price, and at half the ruckus.

Caesium 55

Name: From the Latin 'caesius' meaning Sky blue

Subject Notes:

Caesium is the most volatile of the Alkaline Metal family and indeed one of the most reactive Elements in the realm. She has a large structure and this is ultimately what leads her to be so very reactive. Not having a firm grip on her outer electron means she is overcome by the need to displace it onto any one who may be passing by. She is caustic and oxidises in air (producing a very dangerous superoxide), reacts violently in water and can even dissolve glass. In the event of encountering water Caesium reacts much like the witch in L. Frank Baum's 'Wizard of OZ' fizzing, bubbling and exploding in what could be likened to a human fit or seizure. Eventually she spontaneously ignites (due to the releasing of Hydrogen gas during the reaction).

Because of this, like many of her wayward sisters, she must be locked up in the 'Asylum for Electron challenged Elements' and stored in oil to calm her sufficiently. Thankfully her incarceration in the Asylum makes this psychotic Element relatively rare, so encounters with her are limited to reactors and laboratories.

Subject History:

Subject 55 is very similar to Potassium in her chemistry, and though Caesium has no biological role in humans she can partly replace Potassium in the body. Be that as it may, an overdose would be fatal.

All good lunatics are in touch with time, usually via the phases of the moon. Caesium takes this to a new scientific level. Despite, or perhaps because of her insanity she keeps our world in order with the 'Caesium Clock', responsible for defining the minutes, hours and seconds in the official measurement of time. She can also be used to strengthen the integrity of glass.

Caesium receives little recognition for her good works, and is more often associated with man made disasters related to her radioactive isotope (CS-137) when released in nuclear weapons and reactor meltdowns.

One of the most memorable incidents involving Caesium is that of the Chernobyl disaster in 1986 when the Chernobyl Nuclear Power Plant went into meltdown, and radioactive Caesium escaped and rampaged through Ukraine, Russia and surrounding countries, managing to find her way 1500 miles away from the initial incident. Spanning as far as Great Britain, the irradiated Element carried by the wind and further distributed by the rain affected crops and cattle that grazed on the Caesium laden fields. Contaminating vegetation, meat and dairy and rendering acres of land useless for decades. Back in the homeland of this atrocity animals and children were born deformed, thyroid cancer among children in Belarus, Ukraine and Russia rose sharply and vast forests were wiped out. These are just a few of the effects that Caesium can instigate on escaping her confines.

Cs 55

Caesium

The Metalloids

The Metalloids are a movement rather than a family, an assortment of misfits who have pulled together and linger on the edges of society. The name derives from the Latin word 'metallum' meaning metal and the Greek word 'oeides' meaning resembling in form or appearance'.

As their name suggest they are neither Metal nor Non-Metal but some kind of hybrid in-between: exhibiting characteristics from both. Unlike many groups we will investigate, it is their differences that have gathered together these Elements rather than their similarities.

While they all tend to be semi-conductors and exhibit schizophrenic behaviour, their melting and boiling points and densities are varied, whilst some are toxic others are harmless.

The Metalloids are a well educated, high brow, blue stocking Elements, thus have developed traits far beyond their time. It appears they have developed the ability to change what form they take and even how reactive they become depending with whom they associate. For example Miss Boron, if with Sodium will carry herself as a Non-metal. Curiously however if she interacts with Fluorine, she will engage in Metal like behaviour.

Because of their differences they are somewhat hard to characterize as a group. It is far better to study each subject as an individual case.

Boron 5

Name: From the Arabic 'buraq' which was the name for the mineral borax.

Subject Notes:

Generous and giving Boron is rarely found alone, but in compounds helps to improve the performance of others. For example under Boron's stable and firm supervision the great titans of industry can become even more proficient: Aluminium conductivity is increased, Nickel is easier to refine and Iron finds it easier to flow.

Being a social Element these Boride collaborations also encourage Boron to conduct electricity with more competence than she does as a pure metal.

This high brow Element is also very sturdy, able to withstand the greatest of temperatures and is hard like diamond. Furthermore, like diamond, she is glamorous, tough and has a sparkling personality.

Like many of the Metalloid family Boron is adaptable, taking on different forms: a dark crystal-like embodiment, a metal or a black powder. Ever resilient her powdered appearance is underactive to the wiles of Oxygen, water, acids and alkaline. As previously stated she will react with other metals to improve their performance.

Subject History:

This modest, friendly and strong Element has a sustained an understated history.

In the Ancient world borax was thought to come from the mysterious mountains of Tibet and has enjoyed having her name mentioned in Roman histories and even in Chaucer's 'Canterbury Tales'. She has worked both with the lowliest of women and with Queens as great as Elizabeth I. Together with Lead and Mercury she made up face whitening skin cream, used by hundreds for years despite its toxic effect. Queen Elizabeth was well known for her use of this cream and became ever more reliant on it as she aged.

As times progressed so did Boron. Never one to be left behind she has become the unsung hero of industry. Because of her ability to withstand high temperatures she is used in turbine blades, rocket nozzles and high temperature reaction vessels. With the help of Sodium she toughens heat resistant glass and exhibits her artistic side by helping make ceramic glazes and kitchen equipment.

B ^{5}

Boron

Silicon 14
Name: From the Latin 'Silics' meaning Flint

Subject Notes:
Silicon wants four more electrons and is thus willing to form alliances with most Elements and happy to share as long as she gets her quota. As with many Elements it is this hunger for electrons that makes her so useful to us. She is not afraid to work hard, is well grounded and chemically versatile (much like her sister Element Carbon). This earthly bound Element is also a very fine semi-conductor which makes her perfect for electronics: a huge factor in her success. Her labour comes cheap as she makes herself so abundant in the Earth's crust, second only to Oxygen.
She is responsible for the mundane such as sand and flint but also for the luxurious like Opals, Agate, Rhine Stone and Amethyst. The evidence suggests that though she is a successful working class Element, she has a flair for upper-class trappings and a penchant for the beautiful and desirable.

Subject History:
When one thinks of Silicon many minds will take a metaphorical leap to the cosmetic surgery of the less well endowed. This, however, is a very unfair assessment of Silicon's abilities and history.
Because Subject 14 forms flint she is responsible for the first man made tools and weapons. Without her we would be without glass, concrete or the finest steel to mention but a few.
She is integral to many species on our planet and is believed to be fundamental to our bone growth. With this and her resemblance to Carbon many have come to the conclusion that there may be Silicon based life forms somewhere in the universe. However this theory is flawed. Unlike Carbon, Silicon is typically found as a solid and only turns to gas at very high temperatures ($2355°C$) which are not conducive to life. Additionally breathing Silicon is harmful. Those unlucky enough to have constant exposure to this Element, such as miners or stone cutters, can develop a condition known as 'Silicoses', which results in wheezing, coughing, shortness of breath, and in extreme cases, cancer.
Her resourceful, enterprising attitude combined with her astounding work ethic has assured her of continued good prospects well into the future.
It seems Silicon always has been and always will be one of the foundations of our human life.

14

SI

Silicon

Germanium 32

Name: Patriotically named after the "discovering" country Germany.

Subject Notes:

Miss Germaninum is considered to be the reject of her family. All her sisters are famous (Carbon, Silicon, Tin, and Lead); while Germanium is often overlooked despite her adaptable, hard working nature. Like her sisters she wants her electron and believes in hard work to get it. In addition to this quality, though stern, sullen and some what gloomy, she is a completely stable Element in both air, water and remains unaffected by alkali and acid, with one exception: Nitric Acid.

In spite of all these stoic traits time and again she has been omitted and replaced by other Elements. This has earned her a reputation of been 'unlucky' and seemingly of having no hope, making her bitter, cynical and brittle. The major factor for constant rejection is, though hard working, she is not an accessible Element. She is less abundant than any of her family, and difficult to work with when eventually found. It is these circumstances that persuaded this detached Element to join the Metalloid Revolution.

Subject History:

Mendeleev predicted her existence in 1871 giving her the name 'ekasilicon' (like Silicon). This seemed to set the tone for her future living in Silicon's shadow.

Her first job was working with transistors, because of her genius ability to conduct electricity one way and not the reverse. At first this was a great success. However after Germanium did all the hard work she was soon replaced with her more predictable, cheap, and more abundant sister: Silicon.

Germanium nevertheless does not give up easily and in 1980 picked herself up, dusted herself down and got on with her next job as a health supplement. She supposedly had the ability to improve the immune system, help the human body be more efficient with its Oxygen supply, and generally make one more happy and healthy. This had never been scientifically proven, and in 1982 the British government deemed she had no medical value.

She has never been known to kill, and possesses few health risks. No doubt this world weary Element will be seen again trying to make people observe her worth, and get the recognition she feels she deserves.

32

GE

Germanium

Arsenic 33

Name: Originally from the Greek word 'arsenikon' for yellow 'Orpiment'. Also has links to 'masculine' or 'potent' in Latin.

Subject Notes:

Arsenic is one of the most notorious Elements in the Periodic world and has been around for so long, been in so many places and done so many things that any job Subject 33 now takes on is essentially to distract her from life's more mundane moments. She is self assured, verging on the egotistical, but makes no excuses for what she is and has no reguard for contemporary standards. She shows evidence of a somewhat manic and unstable nature from depressions to passions.

She has great ingenuity and technical skill regarding the art of murder and is often employed for such purposes. One of the numerous reasons Arsenic has become so prolific at her art is that she is a master of disguise and thus discreet. White Arsenic (ASO3) is her most illustrious guise, in which she takes on the form of a white powder with no odour or taste. In this shroud she can easily be slipped into tea, foods or 'medicines' with no evidence of her presence until it is too late. She has often been mistaken for sugar or flour with equally devastating effects. Should this guise fail she can fall back on her other forms: grey, black or yellow Arsenic.

Subject 33 draws pleasure from baffling police inspectors as she is almost impossible to detect. Her killing method makes it easy to disguise a murder as an illness as the symptoms are often the same i.e. stomach cramps, clamminess, cold sweats, coughs, vomiting, colic, and coma. This skill has made her a favorite in the underworld of crime and has earned her the title of 'Inheritance Powder' and 'The Poison of Kings and the King of Poisons'

Although Arsenic has been known and revered from ancient times her real fame started in the Victorian era where she was used in everything from cleaning substances, make up for whitening the face, wallpaper, artificial flowers, a variety of green and yellow paints, rat poison to weed killer, and in some cases even so called miracle cures. It is, however, her killings that brought her real fame.

One of the major investigations concerning this Element is the killings of Mary Ann Cotton (1832-1873). Using Arsenic she killed 16 family members and maybe up to 21 people before being eventually caught and hanged. The victims included her mother, husbands, step-children and even her own children.

For all her dark deeds Sulphates of Arsenic are vital to human life, maintaining a healthy metabolism and production of red blood cells. She has also been used in a fair few medicines both historically and up to this date. But whether that is enough to outweigh all the horrors she has committed in the name of boredom is still to be determined: it is likely that Arsenic does not care either way.

Though Arsenic is no longer used and has been omitted from most products she is still just as infamous as she has ever been.

33

As

Arsenic

Antimony 51

Name: From the Greek 'anti-monos' meaning 'not alone'. Chemical symbol from the Latin 'stibium' the ancient name for Antimony sulfide.

Subject Notes:

Antimony is the image of luxury, over indulgence and the quintessential 'come hither' woman. As her name insinuates Antimony is never alone but often found sponging electrons off others without guilt. The usual suspects who indulge her including Sulphur, Copper, Lead and Silver.

Like all Metalloids Subject 51 is changeable but also frivolous, constantly changing form and shifting properties (a black powder, a lustrous grey metal). Physically she has the traits of her closest friend Sulphur, chemically she acts like a metal, and yet she also resembles her sister Arsenic when it comes to the art of poisoning. This cunning and beguiling Element is just as toxic as Arsenic, though less life threatening. This is due the fact that the body works fast to expel her via vomit or diarrhoea. Once inside the body however Antimony will linger longer than Arsenic and the body has a hard time ridding itself of her. Antimony poisoning symptoms include vomiting, diarrhoea, thirst, kidney and stomach pain, fainting and, if left untreated, death. Subject 51 is infinitely more devious than Arsenic. She takes her time when committing murders. As stated in the beginning of this file a large dose of this Element will simply cause vomiting, but small individual doses subtly kill the victim, starving the body of other vital Elements. Perhaps the most famous suspected murder by Antimony is that of Wolfgang Amadeus Mozart.

Subject History:

Antimony has long been held in awe, fear, and adoration by the human race and has gained herself a dark reputation. She was important in alchemy for various reasons and this linked her in the minds of many to witchcraft and the dark arts. Medieval monks' believed that she had forbidden sexual powers and saw her as a symbol of femininity with her seductive ways and changeable nature.

Her associations with the 'evil woman' go even further back in time. Many cultures used Antimony in black eye make-up or kohl. It is interesting to note that the Bible often links dark eye make up to devilish women; Jezebel is the first that may jump to mind. Even in this modern era black eye liner is associated with witches and dark and mysterious women, a link most likely stemming from Antimony's legacy.

In the introduction to this Element it was said that she was a symbol of over indulgence. This comes from a Roman tradition. At feasts the wealthy would stuff themselves until they could eat and drink no more. Then they would drink from an Antimony lined cup causing them to vomit up the contents of their stomachs, making room for more food and merriment. The Victorians had similar uses for the Element and used her in many medical treatments, including a Pellet of Antimony used for constipation which could be retrieved after the bowels expelled it to be used again and again.

51

SB

Antimony

Tellurium 52
Name: from 'tellus', meaning Earth

Subject Notes:
Tellurium is one of the lesser known Elements. This comes down to three main reasons; she is relatively rare, we have few uses for her and she usually spends her time with more famous Elements that eclipse her presence. Through studying her habits I have deduced that this does not affect her confidence or moral in any way.

Subject 52 is good at heart but possesses a wicked sense of humour, impish streak, and is highly mischievous. Her most notable form of jest is causing extremely bad breath in humans. This condition can last up to 30 hours, and in some men her effects have been known to last up to 8 months. Vitamin C has proved the best remedy for this 'garlic breath'.

Evidence suggests that she is easily influenced by other Elements, and not always for the better. For instance she is not a particularly toxic Element, but when in league with Oxygen and Sodium (Na_2TeO_3) she becomes fatal, causing vomiting, inflammation of the gut, internal bleeding and respiratory failure.

Subject 52 is liked by most Elements, and they find it easy to work with her. She has no ambition or wish to be in the limelight unlike many of the Elements, and is happy to shuffle through life unnoticed and laughing at the misfortune of others.

Subject History:
This impish character proved frustrating and perplexing from the start and was given the names 'metallum problematum' or 'allurm paradoxum', as efforts were made to isolate her and find out her true nature. Eventually in 1796 Franz Joseph Muller von Reichenstien (1740-1825) named her: Tellurium.

As stated above Element 52 is easily influenced by others, but not all her associations are for the worse. Her more honourable friendships include those with Copper – who helps her to become more workable – and with Lead – in whose presence she toughens up and resists acids. She is misguided, but with the right influences can be a very generous and caring Element.

In the human world we use her in stainless steel to make it easier to work with, as a blue and green colouring agent in ceramics, in photoreceptors and micro electric devices with Bismuth. But in general we have little demand for this Element.

52

TE

Tellurium

YOUR
ELEMENT
NEEDS YOU!
Show your support buy a button today!

The Noble Gases – Group 0

As their name suggest the Noble gasses are of an affluent and privileged stock. They reside on the eastern edge of the periodic table where they remain detached from the inferior common Elements, for which they feel a great deal of distain.

These aristocrats are electron rich, having no desire to gain, lose or share any electrons, and believe such mundane squabbles are for lesser Elements. This makes for a very anti-social family; not reacting with other Elements or even socializing with their family, preferring their own company and the solitude of their estates.

There is little to say about these elusive Elements, whose own condescending and aloof natures prevent too close an investigation.

The way in which the majority of these Elements were identified is worth note. When subjected to an electrical current each gas displays a coloured glow unique to that Element, much like a human finger print. New noble gasses were identified by the application of electrical current.

Thus explaining why they are usually employed in lighting, as will be revealed in deeper investigation of the Noble Gas.

Helium 2

Name: From the Greek 'helios' meaning sun.

Subject Notes:

Sister Helium is the Elemental ideal, complete and holy: the purest of all Elements. She is fulfilled, at peace with her existence, and needs not taint herself with other Elements, or by playing their underhanded games in the pursuit of electrons. She cannot be corrupted or altered by any typical means.

She also exhibits extraordinary anomalies that run contrary to all we know of basic chemistry and science. She is constantly attempting to escape to the heavens. Unlike other Elements she is light and small enough to escape through our Ozone layer, ascending into space.

When forced into a liquid form she expands on cooling. We have named this form 'Helium II'. In this form she becomes a million times more conductive thermally than regular liquid Helium, and the peculiarities do not stop here. Nothing can stop her flow- even when contained by regular means she can defy gravity and crawl up the sides of her container in a bid to escape.

She is very much in charge of the Noble Gases family, keeping a close eye on their activities and keeping them in check. Obscurely she seems fond of Madam Neon, this seems strange considering the holiness of Sister Helium and the degenerate pleasures Neon offers. One can speculate that she may see Neon as a project to cleanse, or maybe she secretly delights in the sins of Neon.

Subject History:

Her presence was discovered in the sun long before on earth, as if prophesying the holy and pure nature of her being. The discovery of Helium in the sun's spectrum also opened doors for many other Elements to be discovered such as Caesium, Rubidium and Thallium.

When individuals think of Helium many think of floating balloons for children, however her capabilities go far beyond that. Uses for her include providing inert atmospheres for experimentation (a common job for the Noble Gases), in aircraft, rocket launchers, deep sea diving and Helium-Neon lasers (used in supermarket scanners).

²HE

Helium

Neon

Name: From the Greek 'neos' meaning new.

Subject Notes:

Neon is renowned for her intense crimson lighting and monopolizes the entertainment industry in both the Elemental world and our own. She promises instant gratification and a scandalous time. Like a moth to a flame she lures people into her various amusements.

Even with her love for fun, attention and the sound of the well-to-do falling into disrepute, she cannot be enticed by the vices she (who are generally shy and would rather stay in the background) her disdain for others is the same. In fact, she will not bond with any Element at all. She evidently prefers to remain the ring master of her entertainments and fancies herself above the common Elements, keeping a firm grip on her electrons. Something she can achieve with relative ease as she is a small Element, thus keeping her electrons close to her nucleus.

Subject History:

Madam Neon will not be ignored; she will shine brighter than any other light, can be seen on bright sunny days, over long distances, through fog and smog or city life. It is this quality that makes her so adept at her endeavours. These endeavours include beacon lights, lasers, driving equipment and most famously in advertising. She was first put to use in advertising in 1910 in a Paris motor show, then made the move over to the Americas in 1925 where she became a symbol for the new and exciting. This garish and robust Element is responsible for the increasingly extravagant lights of Las Vegas, late night kebab and takeaway signs, houses of ill repute in the 'red light district' and many other establishments corrosive to the soul.

Further investigation has uncovered that only red lights are pure Neon, other colours classed as 'Neon lights' are a concoction of Elements that she has coerced into playing her games.

N

Ne 10

Neon

Argon 18

Name: from Greek word meaning 'lazy', 'idle' or 'inactive one'

Subject Notes:

Lady Argon's name says it all; she will not lift a finger for anything or any one and is increasingly egocentric. As one of the Noble Gases she is rich in electrons, is non-reactive and needs nothing from any one. All her shells are complete so she needs no interaction with the common folk in the fight for electrons. A truth she is grateful for, having repulsion for the lower classes or self made powers who work for their electrons like the Non-Metals.

By no means is she rare. She is one of the most relatively abundant Elements in the atmosphere, making up 1% of air.

Subject History:

Like many aristocratic families, just because she is not common, does not make her particularly useful. She is not needed for human, animal or plant life. However her ladyship does have attributes that can be useful. She can be used as a blanket gas in experiments, to create an environment where only the Elements one wants to react will react. The best examples of this are in welding and in filament light bulbs. Keeping other Elements in order is her main role, one she takes on with indifference and negligence. Luckily her presence is enough to ensure the Elements behave, and does not often require her attention.

Argon was the first of the Noble Gases to be found, and thus acted as a catalyst to discover her other sisters. The assumption being that if there was one Noble Gas, there was bound to be others.

AR 18

Argon

Krypton 36

Name: From the Greek 'Kryptos' meaning Hidden One.

Subject Notes:

Krypton is a quiet, shy, and harmless Element (unless of course when causing asphyxiation, but the same could be said of most gas based Elements). Miss Krypton may be one of the rarest Elements in our atmosphere, and prefers to stay hidden in the cocktail of gases, rendering her name rather apt.

Subject History:

When the Noble Gas family was discovered, the architect of the periodic table, Mendeleev, refused to believe they were true Elements and initially intended to exclude them from his table.

However the discovery of this Element helped him decided that his table could be modified to include this anti-social family, which was, by this point too large to ignore or count as an anomaly.

As we have seen Noble gases prefer to avoid work wherever possible; Krypton is no exception to this aristocratic tradition. When inclined, however, she does have her uses. Krypton emits a soft violent glow when subjected to an electrical charge, making her ideal for use in lighting strips. It should be noted that Subject 36 is rather ardent when it comes to electronic currents, responding in record time. She deeply dislikes tardiness or those who deem it acceptable to be 'fashionably late'. In her eyes there are no excuses for been late.

In the cold war (1950-1990) she also had a role to play. Krypton-85 is radioactive and given off by some nuclear reactors and by monitoring the amount of radioactive Krypton emanating from the Soviet Block, they could ascertain the extent of the Soviet nuclear production and materials.

36

KR

Krypton

Xenon 54

Name: from Greek 'Xenos' meaning 'Stranger'

Subject Notes:

As her class suggests Xenon is another of the inactive gases, she is insular, elusive and fleeting in nature. Even in the most important of social outings, when the most anti-social of us feel obliged to attend, Xenon is far from the public eye. Rather than trouble herself with the obligations of a well-to-do family, you would find her lurking in the libraries or studies, enjoying her lonesome sanctuary.

As was made clear at the beginning of the chapter, the Noble Gases do not react with other Elements, that is, unless humans interfere and expose them to harsh environments: in this case sub zero temperatures.

Xenon is relatively easy to persuade, bonding at -40°c whilst under pressure, a relative sunny day compared to the absolute zero and ultraviolet radiation it takes to get stubborn Argon to bond. Xenon is easily lured into bonding, compared to her sisters, because of her large atomic structure. Her outer electrons are further away from her nucleus, so she has less of a grip on them, making it easier to pry them away from her. Just as the Alkaline Metals become more reactive as one looks down the group (or as their atomic mass increases), the same can be said for the Noble Gases.

Subject History:

The elusive Xenon was the last of the Noble Gases to be discovered (save Radon), as she skulked in the metaphorical library of the ether, struggling to remain anonymous. It was for this reason she was given her name: 'The Stranger'.

She was eventually identified by the same technique as her sisters, by applying an electric current and pinpointing her colour finger print - a disarming pale blue It is also for this property that we utilize her in camera flashes, head lights in new cars and other specialized lamps. Occasionally she is made useful in creating a non-reactive environment for experiments, but as far as Elements go we have very little use for her. In the 1950's doctors and scientists dabbled in using her as a new anaesthetic gas for surgery and she proved to be a far better option than the present gas (Nitrous Oxide) as using Xenon had much fewer side effects. However, as so many Elements in this society, she proved too expensive to be viable. Thus today our main use for her is in lighting.

She remains happy with her small roles, keeping herself out of view of the prying public and giving her time to pursue her own pleasure whatever they maybe.

The Post Transition Metals

The Post Transition Metals are also named 'The Poor Metals', they are outcasts from the Transition Metals, coming together to form their own family. But in what sense are they poor? Investigations into these Elements have shown it is not from lack of wealth or fame (indeed some of these Ladies have made quite a name for themselves, though not always for commendable reasons), but rather because they make poor metals. They are softer, weaker, less reactive, and their boiling and melting points are considerably lower than the stereotypical Transition metal.

Despite the fact they have chosen not to be associated with the larger group they are definitely metals by nature and could not readily be mistaken for one of the women in the Metalloid movement.

Aluminium 13
Name: from the Latin name 'alumen' or 'alum' meaning bitter salt

Subject Notes:
It could be said that we live in the age of Aluminium. No matter how impoverished or wealthy, how young or old we all use the industrious Aluminium in our day-to-day lives, from the ordinary, such as cooking foil, door handles, tubes, cables, and cans to the extraordinary like: aeroplanes, boats, reflecting blankets, and solar mirrors. These are just a few of her many enterprises.

She possesses many qualities that make her instrumental in our lives. Firstly she is an excellent conductor of both electricity and heat, making her adept for cables and the likes.

Secondly unlike Iron (her industrial rival) she is light weight, easy to use, not to mention cheap to synthesize and utilize.

Thirdly she is a very exuberant Element. Aluminium is the most abundant metal in the earth's crust making up 8% of it and she is a favourite in industry thanks to her being economical and easy to recycle.

Lastly she is a very capable lady, and can look after herself. She is arduous to ignite and protects herself with an oxide layer preventing rust and corrosion. This makes her goods long lasting and tough. All together Ms Aluminium is a very useful Element to have around.

However Aluminium is not needed in a biological sense to our human form. Our bodies do not readily absorb her, but if they do it is very hard to get her out of our systems again. In small doses she will not harm us, but an overdose could be fatal as once in the blood stream she will work her way up to the brain were she wreaks havoc, causing brain damage and has being linked to Alzheimer's.

Subject History:
Aluminium was involved in the first true chemical industry in our world.

In the beginning she was a lowly fixer of dyes, and paper preservatives in the ancient world and was in a salve to stop the bleeding of small cuts and wounds. In 1825 she was isolated as her own metal, but the process was hard, time consuming and expensive, so only the rich could afford her services. Emperor Napoleon III had cutlery made by her to show off his wealth and forward thinking. Many people looked at and lusted after her products, but because of the cost most could not afford them.

This did not please the ever ambitious of Ms Aluminium, who wanted to expand her empire further. In 1886 a cheaper more viable way of extracting the metal was found, after which the price of Aluminium dropped to one thousandth of the previous price. This opened doors to a wider audience than ever. Subject 13 still had a stumbling block in her dream of domination of the industrial world. The main mines were in Greece and Italy which gave the Pope monopoly over manufacture in Europe. But Aluminium is always finding ways to better herself and push forward.

Mines were found in north Yorkshire, pushing the price down yet again. Taking her self from the rich man's metal to a metal for the people.

13

AL

Aluminium

Gallium 31

Name: Derived from the Latin name for France "Gallio"

Subject Notes:

Subject 31 is liquid at around room temperature (30°c to be precise). She boasts the largest range of temperatures at which she remains in that state, showing a fondness for it. Unlike the majority of metals Gallium will shrink in volume when melting. Interestingly the only other Elements to posses this quality are Antimony and Bismuth. Miss Gallium is of the opinion that any problem can be solved by a good cup of tea, and thus delights in hosting a multitude of tea parties. In fact she attends and throws so many tea parties that despite being a rather common Element (more so than Lead) she is difficult to locate as she is widely dispersed throughout the Realm, attending social events.

These tea parties also offer her the perfect opportunity to perform her world famous parlour trick: that of the disappearing spoon. When in solid state Gallium resembles Aluminium, thus a spoon fashioned in this Element looks like any other, but when one comes to stir their beverage, the spoon 'disappears' in the hot tea, much to the hilarity and delight of her guests.

Subject 31 is repelled by all who frequent 'The Asylum for Electron Challenged Elements' especially the Halogen and Alkaline Metals finding them to be especially repugnant. Their constant faux pas and contemptuous behaviour vexes her and offends her feminine sensibilities and she will react with them and in some case even dissolve or faint. She is much better suited to dining with other members of her family, in particular Indium her best friend, and companion in gossip.

Subject History:

Gallium is a mild mannered Element, being low in toxicity and has very few health risks. Though she can slightly stimulate the metabolism of our humanoid form, she has no vital role to play or long lasting benefits, freeing her time up for matters more important to her, like afternoon tea.

She is never usually mined just for herself, and is merely a by-product of other Elements. Be that as it may, we have found momentous jobs for her to take on. Gallium is a semi-conductor, and has replaced Silicon in some supercomputers and mobile phones as she is calmer and generates less heat, making her more suitable for the job.

GA ^31

Gallium

Indium 49

Name: from Latin 'indium' meaning violet or indigo

Subject Notes:

Indium is a Lady of Leisure, and avoids hard work where possible. She is far more interested in her outward appearance. So fanatical is she with this aspect of grooming she will bind her self to glass, until she evaporates leaving a mirror like residue. The reflection is equal to that of silver, but Indium is not so susceptible to oxidization or corrosion, ensuring she can always see her own image looking back at her.

Other than this forgivable eccentricity she is a stable Element, air or water having no effect on her constitution (although acids have an undesirable effect). It is worth noting however that she has a low threshold for pain, even bending her will elicit a high pitched shriek.

Subject History:

As stated above Lady Indium dislikes work and we have few uses for her in our human world. What uses we do have for her stem from a characteristic of all English women of fine breading: an intolerance for heat. It does not take much to reduce her to a pool of shiny liquid. Because of this she is used in low melting alloys in fire sprinklers in warehouses. She will also now and again try her hand at being a semi-conductor in electronics. She is, nevertheless, usually just a by-product of smelting Zinc or Lead Sulphates.

She has no biological role either, though tiny amounts may stimulate the metabolism. However one should be careful engaging in this practise, for in excessive doses she will set off a toxic reaction affecting major organs including the liver, heart and kidneys.

IN 49

Indium

Tin 50

Name: The symbol from the Latin "stannum" meaning Tin. Tin is the Anglo-Saxon word for the metal.

Subject Notes:

Tin is an honest, hard working Element. The high born have their Gold, Platinum and Silver; other metals live in the domain of the blacksmith. But Tin is the every man's metal, the working class hero. Soft and malleable enough to form with a simple hammer, but strong enough to be useful for everyday trinkets and utensils. Not only is she cheap and compliant but she can be melted down and recast over and over, like a phoenix she is a sign of hope for the downhearted and broken: one can always rise from the ashes.

With a shining smile she can resist air and water, and protects exposed surfaces with an oxide layer. Though she has a cheerful and determined disposition she is a sensitive soul at heart and when bent she will cry out. Interestingly when temperatures become too cold she will suffer from 'Tin Plague'. This is when, because of the cold, she slowly disintegrates, turning to nothing more than Tin dust.

Subject History:

She may be humble and puts on no airs and graces, but this does not mean her contribution to history is small.

The ancients knew of her and she has been written into history in the form of trinkets found with the Pharaohs of Egypt and on Machu Picchu from the Incas.

Arguably it was her partnership with Copper that really kick started her career. Added to Copper the result is bronze, which had a whole age named after it, such was their contribution to society. Bronze is easier to work with than earlier metals and much harder, making it perfect for weapons, jewellery and tools.

There are of course draw backs to Tin's willingness to work with anybody. For example Tin worked with a plethora of other Elements to create pewter, which was very useful for cheap cups and plates, but also toxic thanks to the presence of Lead and it soon fell out of favour.

Today we use her alloys just as much as we ever have. From billions of food cans to solder, and metal used in bells to dental amalgams. These are just a few of her uses and a tiny note from her history. The events and accomplishments of this Element could easily have filled their own book.

50

SN

Tin

Thallium 81

Name: from the Greek word 'thallos' meaning green shoots or twig, because of its spectral colour.

Subject Notes:

Thallium is the most poisonous Element yet to be discovered. She has made an art of death and has a long and gruesome resume. What makes her such a diabolical Element compared to the other killers? Other toxic Elements, like Fluorine are so reactive they are rarely found in nature alone and cause less damage bonded up, and Antimony is so poisonous the body immediately reacts to eject her from the body. But Thallium sits and waits patiently for unexpecting passers by to ingest her deadly toxin. Once inside the body her true genius comes into play. Because of her electron configuration she can happily give or take a number of electrons to mimic other Elements that the body needs (namely Potassium). Once inside the body she strips the pretence and her true identity is revealed, then nothing can be done. She is free to roam at will causing havoc wherever she wishes, smashing amino acids and breaking cells. As the body realises its mistake it tries to expel her via the intestines, only to make the same error: reabsorbing her into the body, once again mistaken for Potassium.

 She is the queen of Poisons, and has truly made it an artistic profession.

Death by Thallium is a very slow and excruciating experience as she draws out her art. Symptoms may include: loss of hair, loss of finger nails, lethargy, numbness, tingling in the hands and feet, blackouts, slurred speech and damage to peripheral nerves. Victims may experience a sensation of walking on hot coals.

Subject History:

For a long period of time Thallium poisoning was thought to be incurable. Who would have thought such a simple household staple like Prussian blue ink could thwart such a mastermind? Prussian blue ink contains high volumes of Potassium which, once in the body, displaces the Thallium and has a much stronger grip. Thallium was first used as a treatment for ring worm, not that she actually cured ringworm, just made the patient lose their hair making it easier to treat. Once it was known that this Element caused baldness this became her primary use, up till the point when it was discovered that she could kill. She then took up a rather more sinister job.

Her killings earned her a place in Agatha Christy's novel 'The Pale Horse' where she came to be better known to the public, causing and preventing a few murders as a result. Thallium was even considered as the poison to kill Castro, though this idea was abandoned. Others though were not so lucky. Despite her assassination contracts, many others have killed themselves and others without their knowledge. This just goes to prove how dangerous and treacherous this Element can be.

81

Thallium

Tl

Lead 82

Name: The origin of this name is unknown. However the Romans knew her as 'plumbum' and this is how she obtained her chemical symbol Pb.

Subject Notes:

Lead, the Element that promises so much, that rose so high only to deliver death, be rejected and scorned. This dull and melancholy Element now lives in the memory of past glories reduced to menial tasks, when once she served kings. She is soft, a slave to her emotions, easily affected, tarnishing in air and has a tendency to 'creep' (see glossary) under the slightest pressure such as gravity.

One may have heard of such phrases as 'as heavy as Lead' and 'a heart of Lead'. This is not just because of her substantial weight but because such is the air of misery surrounding her that she can bring down the most cheerful of souls.

However one should not dismiss the qualities that let her rise to fame before her dark side was discovered. Firstly she is slow to corrode and is perfectly stable in water and air, though a little dulling does take place. Secondly she is ductile, malleable, easy to work with and is excellent at preserving various items.

Subject History:

Lead was one of the most economically important Elements there was. In fact many have a theory that it was thanks to her that the Roman and British Empires could spread so far, and last so long. She was used for the cups of Kings and Emperors, the cosmetics of Ladies and Queens, vibrant pottery glazes and paints, from toys to bullets and an array of other trinkets, bringing water to the rich and poor alike in pipes and aqueduct linings, pewter pots and cups. She was held in high regard for her skills in telling the future in many different cultures and was one of the base metals of Alchemy. She was even associated with the Roman god Saturn, ironically the God of melancholy.

For all that, after years of living the high life and basking in fame she was to fall from grace. Lead is now known as the Element that kills, drives Kings mad, and causes sickness. Some have reason to believe it was because of her that the Roman Empire fell. Emperors who most likely had ready access to Lead had a reputation for madness and paranoia, as well as seeming to be rather unlucky in fathering children – all symptoms of Lead poisoning. Even now she is still associated with death. For hundreds of years she was utilized in coffins, partly because of her ability to preserve the body after death. Parents fear paint and painted toys; house owners dread old buildings in case they still hold Lead piping or window frames.

Lead's day has passed. Perhaps then it is no wonder she lives in perpetual gloom, grave in spirit, constantly seeking fame lost and committing murders out of spite or revenge, preying on those unfortunate enough to forget her deadly skills.

82

PB

Lead

Bismuth 83

Name: From 'Bisemutum' the Latinized version of the old German word 'weissmuth' meaning white substance.

Subject Notes:

Bismuth is a simple peddler of goods and is something of an anathema to other more prim Elements. This is because she dwells in the south-east corner of the Periodic Table, which is reputed for its poison and bête noire Elements (Thallium, Lead, and Antimony). So toxic is this shady area that some have named it 'Prisoners Row'. Bismuth surrounds herself in every direction with insidious characters famed for their brazen killing and disregard for life, which in turn has given her a bad reputation: Many deeming her guilty by association. On closer inspection we can see nothing could be further from the truth. Subject 83 is the largest stable Element and so is not burdened by an explosive temper, and rather than being toxic she dedicates a portion of her time to curing the sick (discussed in Subject History).

As well as associating with wicked Elements she also displays a few other eccentricities, namely expanding like ice on freezing, contrary to accepted behaviour on solidifying. This is not the end of this mysterious and beautiful Element's talents, for on cooling she forms splendid 'Hopper crystals' for which she is well known; Petrol coloured, Escher like sculptures which she sells from the back of her caravan.

Element 83 is the most naturally diamagnetic metal, and interestingly is repelled by any magnetic field.

Subject History:

As stated in Subject Notes this Element is of some medical use. Bismuth subnitrate and subcarbonate are used to treat stomach upsets, extreme bouts of uncontrolled defecation, and gastric ulcers.

On her travels she has also picked up a number of other skills that have given her the potential to trade and flourish. For example we use her in many cosmetics like lipstick and nail varnish to give it a pearl like shine, in an array of yellow and red paints; thanks to her low melting point we can utilize her in fire detection and extinguishing, , and she also replaced Lead in some bullets.

Bi 83

Bismuth

The Halogen Group VII

The Halogens are an extremely volatile, unhinged and deranged Phylum and are best incarcerated behind the walls of 'The Asylum for Electron Challenged Elements'. For all are poisonous and are responsible for a history of savage murders, frenzied attacks and nefarious maiming. Should a member of this family ever escape from the 'Asylum' pandemonium and bedlam would no-doubt ensue.

When in close proximity to a Halogen one should always take great care. Even when sedated the Halogens are best avoided.

The underlying motivation for this deranged and erratic behaviour is their electron configuration. They all possess 7 electrons in their outer shell, which means they want to fill it by any means possible, and have very little in the way of self control. This makes for some uncontrollable savagery to get their electron fix. As we look down the group the Halogens become less explosive, because their outer shell is further from their nucleus they are less inclined to fight for electrons. Further research reveals that the boiling and melting points of Halogens increase with atomic mass, meaning Elements at the lower end of the periodic table have an infinitesimal amount of sanity in comparison to their sisters. Hence Iodine, the most compos mentis of the family, manages to avoid the confines of the asylum, though for how long is always in debate

Fluorine 9

Name: Named after the ore from whence she came 'fluorspar'. The 'fluo' from the Latin word 'fluere' which means to flow like a current- because that's what she allows metal to do (I.e. to flow from their ores during smelting)

Subject Notes:

Subject 9 is arguably the most reactive of all the Elements, with a volatile, unpredictable, explosive temperament and aggressive nature.

Although keeping Fluorine under lock and key at all times does not eradicate her attempts to escape, it is necessary to ensure she remains within the confines of her cell at 'The Asylum for Electron Challenged Elements'. Patient 9's state is such that she will never be permitted any degree of freedom or exoneration for all know that the results of such liberation would be disastrous.

Her insanity is due to her desperate need for an electron and has very little in the way of shielding and self-control. This deficiency is too strong for her to suppress and she will kill, bite and maim anything standing between her and that elusive electron.

Patient 9 has no comprehension of consequence, no moral understanding and as expected from the 'hysterically insane' no conscience. This tiny ferocious Element is a highly toxic (even her smell will attack the nostrils, a fact she finds delightful and hilarious), corrosive and flammable gas: a terrifying combination. When combined with the fact that she has a somewhat lax comprehension of reality this makes for a most unpleasant Element.

She will ravage most other Elements to steal their precious electrons, with the exception of a handful of the Noble Gases. Argon, Helium and Neon whose firm grasp and unbreakable will make their electrons impossible to pilfer.

Subject History:

Several attempts to isolate and tame this unhinged monstrosity have ended in blindness at best and fatalities at worst. She has proven very difficult to contain and can be devastating in her hydrofluoric state even in small doses. Sir Humphrey Davy wrote this warning in his notes of subject 9:

"Is a very active substance, and must be examined with great caution."

The wise diligently heed this advice.

Despite this some have managed to channel her energy and violence into something useful rather than destructive.

For example when added to water she can use her aggression to help strengthen and partially rebuild ones teeth, as well as help purify water.

She is probably most commonly known for her roles in toothpaste and insecticide. She must, however, be kept under supervision, never trusted on her own and handled with a healthy respect for the damage she can cause in her lust for the electron fix.

F ⁹

Fluorine

Chlorine 17

Name: from Greek 'chloros' meaning greenish-yellow

Subject Notes:

Like all those in the Halogen family, Chlorine is maniacal and aggressive in her natural form. Fortunately she is not usually found this way, though occasionally she slips under the radar to cause havoc.

Her sharp smell will hit you first, rapidly followed by her claws, which she will direct straight at eyes and lungs. Even a short period of exposure to her in this gaseous state is highly damaging. Upon attacks she can turn skin green, yellow or black, make eyes water and turn cataractous and eventually will kill her victims by drowning them as fluid builds up in their lungs.

Chlorine is not particular about where or how she obtains her electrons, will react with all known Elements, save the Noble gases, and gleefully mauls away at metals. For all her flaws Patient 17 can also be tremendously helpful, this observations leads to Chlorine often being diagnosed as schizophrenic. Under the right guidance and electron medication Patient 17 can almost be considered sane, when in this state she is referred to as 'Chloride' (or Cl-, when she has acquired an electron and is stable).

Subject History:

Subject 17 has a long and colourful history, and seems to cause harm and help in equal measure. For every questionable deed one of her personalities commits, the other will match it with one good.

Let us start by investigating the bad. Chlorine was first put to use in World War 1 in the form of mustard gas. Though when harnessed as a weapon she out performed her sister – Bromine, who was easily outwitted by the gas mask. For this reason Chlorine was rendered obsolete by World War 2. When bonded with Hydrogen she forms the famous Hydrochloric acid, the fiend at the heart of many accidents and crimes. This duo however is also the acid in our stomachs that enables us to digest food. She is also vital to human life in that she helps to regulate other body functions. Like Fluorine we can harness Chlorine's willingness to attack anything for beneficial purposes. She is used in household cleaning bleach, for bleaching paper and cloth, and famously in swimming pools. In WW1 Chlorinated water was introduced, which probably saved more lives during the war than the Mustard Gas took. And so her good deeds also become apparent. In summary: to utilize Patient 17 one must know how to control her, give her the right incentive and always be on guard.

Bromine 35
Name: from Latin meaning 'Stench'

Subject Notes:
Inmate 35 is no friendlier than her siblings and has just as much energy to expend in
the name of an electron. She will happily steal it from who ever she can get her dirty
hands on, cleaving weak Elements without thought or care. She has only one goal: a
full shell.
She is one of the few Elements that is liquid at room temperature (Mercury, Gallium
and Caesium being the others), deep red in colour and offensive in smell as her name
suggests.
Though she is cunning, intelligent she is not. What she lacks in intelligence, however, she makes up for in brutality. That said she is not quite as deadly as her previously discussed sisters, but should nevertheless not be underestimated at any point.

Subject History:
Bromine had links to royalty through the purple dye 'Murex brandaris' or 'Tyrian
Purple'. This was used for the togas worn by Roman emperors and was incredibly expensive. In fact the bible mentions merchants that sell rubies, coral, emeralds along
with purple dyes, showing how valuable this product was.
As we moved into the 19th Century Bromine became a Victorian tranquilizer, Patient
35 depresses mental activity and sex drive, somewhat ironic for such an energetic
Element (Though the side effects of such a medication included loss of weight and
depression).
As alluded to in the case file of Chlorine, Bromine was also used as a chemical weapon in the First World War, deadly if , attacking lungs, eyes and eventually killing the
victim. However she was easily thwarted by a gust of wind in the wrong direction,
or as happened in Russia, by extreme cold, causing her to freeze and making her
useless. Needless to say she was soon dismissed as a weapon and replaced by more
aggressive Elements.
Today she is still in the business of killing but her attentions are turned to household
pests. She is rather effective at this as she can dispatch bugs at what ever lifecycle
stage they are at.
 One may think that for such a notorious Element this task must surely seem mundane, but when one considers that she can often be found chasing her own shadow it
seems more reasonable. Bromine delights in exterminating anything no matter how
small. Keeping her distracted in such ways, is the best method for controlling Inmate
35.

Iodine 53
Name: From the Greek 'iodes' meaning violet

Subject Notes:

Iodine is a rather perplexing character, not as psychotic as other members of her family she manages to stay out of the Asylum. She is still an electron thief and desperately wants an electron to make her feel whole but does not display as much violence in order to get it; her methods are altogether more devious. She does however show the same signs of schizophrenia as her lurid sisters.

Like all good women its Iodine's prerogative to change her mind, one never knows how she will receive you. Iodine can be warm, affectionate and tender, soothing your wounds and smothering you in confectionary. She can also cause irritation, stinging and wounds of her own. In such a disposition it is advised that Iodine is kept well away from the eyes. Although in small doses Iodine is not lethal she is still toxic, as one would expect her toxicity increases with quantity. Iodine is intense, often overwhelming and has a flare for melodrama and as such regular exposure can be fatal. That being said she is fundamental to human life, mostly found in the thyroid; she is needed for normal growth and development. A deficiency in Iodine results in various illnesses. One such illness is hypothyroidism, where a person becomes listless and cold. Too much Iodine however is just as bad. Balance and vigilance is required when handling this mischievous and over dramatic character, as is the case with all the Halogen family.

Subject History:

Ms Iodine is a very busy woman and rarely stops, except to indulge in a pot of seaweed tea, of which she is particularly fond. This assiduous Element has had many jobs through history and is no less industrious today.

Miss Potassium seems to bring out the best in Iodine, the dyad are often used to disinfect wounds. Many will remember the dark liquid used on cuts and the tell tale sting that precedes it. Another achievement for good that they have accomplished is in table salt. When added to salt the duo help to prevent thyroid disease.

Iodine had a somewhat more glamorous role during the rule of the Roman Empire, when she was used in the production of purple dye. The colour was obtained from shellfish and seaweed, one drop of dye could be extracted from each shell and to make one gram of the dye eight thousand molluscs had to be used. The tedious nature of the production of the dye made purple an expensive luxury that only the wealthiest could afford. Such Is the quality of Iodine's dyes that they are still used today in inks and printers, though the method of extraction is now somewhat more cost and time effective.

I 53

Iodine

COMING SOON!

THE ILLUSTRATED GUIDE TO THE ELEMENTS

VOLUME II

YOU HAVE DISCOVERED THE FIRST HALF OF THE ELEMENTAL WORLD.

NOW INVESTIGATE THE BIGGEST GROUP: THE TRANSITION METALS, PLUS THE ALKALINE EARTH METALS!

The Non-Metals

The Non-metals are a mixed family consisting of both solids and gases. They tend
to be wealthy and powerful, considering themselves firmly within the upper classes.
Unlike the Noble Gases they did not start out affluent or aristocratic, but made their
fortunes by hard work, well timed and placed Elemental alliances and making themselves invaluable to all.

When one considers the large population of Metals, the Non-metals by comparison
are relatively small. Even when including sub families like the Halogens, Noble Gases
and Metalloids they still form less than a quarter of the total Elements.

Despite this fact they are an elite group within the Elemental world, they make up
most of the earths crust, atmosphere, oceans, and the majority of living organisms.
They are the ruling class among the Elemental and biological worlds.

The reason they have been able to achieve such power is the fact that they are
remarkably electronegative. They can attract electrons powerfully and bend other
Elements to their will once they have power over their electrons, none more so than
Oxygen.

In general they do not possess the same capabilities as the majority of the Metal Elements. That is to say they are poor conductors of heat and do not conduct electricity
(of course there are always exceptions to the rule, in this case it is Madam Carbon in
the form of Graphite).

Hydrogen 1
Name: derived from the Greek 'Hydro' and 'genes' meaning water forming

Subject Notes:
Lady Hydrogen is the simplest of all Elements, having one shell and one electron. She is also one of the most important Elements in existence, holding together strands of life and matter with a vice like grip, a fact that she likes to make sure others know. The thought of one Element having so much power over all life can be more than a little unsettling, only to be intensified when one discovers the peculiarity of this Element. Hydrogen is a simple minded Element and full of her own importance: not an endearing trait no matter how justified. She saunters around in her favourite gaseous form, but when things are not to her liking, such as when other gases try to mix with her or enter her social group without prior consent, she has been known to become explosive and obnoxious.

Element 1 would very much like to have her fill of electrons as is the case with any Element, thus it is common for her to manipulate others with varying degrees of integrity. Results range from the life sustaining water (H_2O) to life denying Prussic acid or Cyanide (HCN).

Subject History:
The arrogance of Subject 1 is astounding considering she does not amount to much on her own. She does however know how to make alliances and business transactions that have elevated her high up in Elemental society.

Hydrogen likes to know about everything going on in the realm, especially if there is electron profit to be made or exploited. The presence of this woman is everywhere, accounting for up to 88% of all atoms in the universe.

Almost all are aware of her part in water, and many know of her harnessing of Chlorine to create Hydrochloric acid, the means by which we digest our food. What many may not realise is that Hydrogen is a component of every molecule and thus every living cell. With her associates Nitrogen and Oxygen she creates DNA. The double helix of DNA is held together by the strong bonds and capable hands of Hydrogen. The same is said for all protein molecules which are the building blocks for organs, muscles, and everything in-between.

These are some of the direct effects she has on our lives, but there are also more subtle but equally important jobs she takes on. One such job is in ice. Ice floats because of the large spacing between her bonds, defying what we know of solids (typically a substance in its solid state is more dense than its liquid equivalent and therefore sinks). This anomaly, that Hydrogen creates with ice means that sea and ocean levels remain stable and the rivers do not flood. It also helps to regulate Earth's temperature and keep it at a viable level for life. She enjoys breaking the rules and revels in the extent of her power that allows her to do anything she wishes with very few consequences.

1

H

Hydrogen

Carbon 6
Name: from Latin: carbo "coal"

Subject Notes:
Madam Carbon can be found everywhere in the Realm. She is probably one of the most socially hard working Elements there is. All social niceties however are purely selfish; Carbon only wants one thing – to get her electron fix. She does not care how or whom she has to be to get it, she is indifferent to others wants or needs and can only see her own electron obsession. Because of this obsession she has an immeasurable talent for being all things to all people. She can be glamorous and impenetrable like diamond, soft and shady like graphite, mysterious and crepuscular like Coal. These are but a few of her guises, she can also hone her talents to match what is needed, for example she can be a good conductor (graphite) or a non-conductor (diamond).

If one had an electron within the vicinity of this woman she would sniff it out and be flattering and fawning within seconds, becoming what you most desired.

She is however not the flirtatious vamp she seems on the surface. The bonds she makes with her clients are stable, and once she has found a source of electrons she will not let go easily, as well as being able to withstand most other chemical attacks that may come her way.

Subject History:
On her own Carbon is not known to be a natural killer, though when bonded to other Elements of a more sinister nature she will bend to peer pressure. In these circumstances there seem to be no bounds to what she will do. For example when she bonds with Oxygen she can become Carbon monoxide (CO), or with Nitrogen and Hydrogen she can become part of the famous killing trio Cyanide (HCN). With these bonds and many others besides she has taken countless lives in the name of an electron.

In the form of coal, diamond and graphite Carbon has always been part of our lives. She had her rise to fame when the making, selling and using of charcoal came into vogue in 30,000 BC when she became and stayed vital to human life.

As we developed we found more and more uses for her, making her indispensable. Now she is one of the most used Elements on the planet.

C⁶

Carbon

Nitrogen 7

Name: derived from the Greek 'nitron' and 'genes' meaning nitre forming. Nitre was the old name for Potassium Nitrate (KNO3)

Subject Notes:

At first glance Miss Nitrogen seems a gentle, well mannered and calm Element. She revels in her reputation as a Lady of fine breeding and in all the luxuries that go with it. Her lack of enthusiasm or will to react makes her too benign to be of any harm or use to us. Like many of the Non-Metals she is a master of picking her friends, and likes multiple bonds to be happy; it is these friendships she forges that have lifted her to an elevated position. Closer inspection of this seemingly quiescent Element reveals a rather more sinister and explosive character. For lurking under that kind and tranquil veneer is a killer of epic proportions.

Some of the carnage created by Element 7 is just a means to an end: she adores being a gas and will fight and tantrum to remain in that form. In doing this she will create a vast amount of heat which causes violent outbursts and explosions of temper. Some of her killings are well thought out however and she knows who to enlist as accomplices. So many poisons, explosives and other generally nasty substances are the workings of this two faced Element. Some of these underground affiliations include the temperamental bond of Nitro-glycerine, TNT, Prussic Acid, Nitrogen Dioxide, Hydrogen Cyanide and Ammonia.

She nevertheless can kill kindly if she takes a liking to her victim. Nitrogen makes up 80% of the air therefore the human body is used to her comings and goings, assuming her intentions are pure. If that 80% should rise even a little she becomes deadly. She will cause no pain and relaxes her victims, sending them into an endless sleep as she starves them of Oxygen. Such is the diversity of this twisted Element that your life and how it ends depends on the whim of this serial killer in disguise.

Subject History:

When in the company of more wholesome Elements Nitrogen can keep the façade of a kindly Lady intact, and indeed does much good for those she nurses. She is needed in DNA, haemoglobin, and is a key player in amino acids that keep life running. Subject 7 is utilized in preserving genetic material, in fertilizers, plastics, nylon, dyes, and anaesthetic laughing gas.

In conclusion, by studying her habits I have uncovered that by day she works in helping and sustaining life, keeping up appearances of the genteel and caring soul. When night comes however, she sneaks out to conduct a plethora of unsavoury acts and murders. It is unclear though if she is conscious of this change in her person or if an alter ego takes over. With the depth of cunning involved in such a masquerade and her legendary bad temper when things do not go her way we can assume she does know and is very good at concealing her true nature. If however she is ignorant of her violent activities, then a place in 'The Asylum for Electron Challenged Elements' should be considered.

N

7

Nitrogen

Oxygen 8

Name: Oxygen from the Greek words 'oxys' meaning acid and 'genes' meaning forming

Subject Notes:

When one thinks of Oxygen the first thoughts are usually of how she is the vital essence of life! She is life giving, refreshing and wonderful. However, when one looks closely her dark side shows its face and contrary facts begin to appear.

It is true that she is essential to all life. We as humans depend on her with every breath and if denied her we would simply die within seconds. Animals and plants are no less dependent on her. Her presence in our rivers, seas and oceans make it so they can team with life. She is vital in the photosynthesis cycle, as well as the CNO cycle (Carbon–Nitrogen–Oxygen*). In short she has made herself indispensible, placing herself at the centre of our lives.

But Lady Oxygen never gives something for nothing and the price we must pay for our addiction is of the highest order. As a candle will burn at a much faster rate in an Oxygen rich atmosphere, so do humans. With every breath we take she ages our bodies and, paradoxically, brings us one step closer to the grave.

Her grip is not lessened in the Elemental domain, if anything she may cause more terror. In A.C. Doyle's The Final Problem Sherlock Holmes describes his nemesis Prof. Moriarty as "organizer of half that is evil and nearly all that is undetected in this great city" this could be said also of Oxygen in the Elemental world. She has some kind of effect on almost all of the Elements. She is the instigator of chaos and corruption, turning the strongest Elements like Iron to nothing but useless flaking rust. She eats away at others, dulling their usually shiny spirits, causing them to make oxidizing layers to protect themselves from her corrosive ways.

She does not rest there; she can also pay off others with electrons to create murderous duos. She manipulates harmless Elements like Madam Carbon into deadly Carbon monoxide (CO). Perhaps the most cunning of her bloodthirsty teams is her alliance with Arsenic. Arsenic on her own is a safe Element to ingest. It is only when she pairs up with Oxygen that she becomes such a toxic fiend. When people talk of the poison Arsenic what they usually mean is the white, tasteless, odourless powder Arsenic Trioxide (AS2O3). It is this powerhouse that makes the subtle killings of Victorian horror. The beauty of Oxygen's plan is that no one suspects her of the evil deed; she can happily cause havoc whilst keeping her 'vital essence' reputation intact. Studying Oxygen has brought me to the rather depressing conclusion that all life, in the Elemental world or otherwise, is a battle against the effects of this Element, whether the aging of our bodies or the rusting of our physical world.

8

Oxygen

Phosphorus 15

Name: From the Greek 'phosphoros' meaning the bringer of light, Also the name for the evening star or Venus.

Subject Notes:

Phosphorus is the Element of gothic horror novels with her ghoulish, eerie glow when in the dark, feeding off the vital electrons of others, and tendency to combust in sunlight; a favourite of poets and writers for centuries. Subject 15 has several forms: white, red and black. Because of her temperamental disposition she is never found alone in nature and for this we should be thankful.

White Phosphorous is the most notorious of her forms and is highly toxic to humans. Once ingested she will immediately set to work destroying the liver; her victim will likely die within the week (not that that has stopped us from trying to make toxic Elements into miracle cures: see Subject History).

As alluded to in the start of this study Element 15 is uncommonly flammable and pyrophoric on contact with air, and once alight this vamparic Element will emit a garlic-like odour.

If exposed to sunlight or heat she will undergo changes that morph her into her second form: red Phosphorous. In this form she is less likely to spontaneously ignite, though is still highly combustible. She will continue to darken in sunlight to more purple hues. Under pressure she will change again to a black, graphite like form and least reactive - the least toxic of all her forms.

Subject History:

This frightening and chilling Element though of a disreputable pedigree is needed for human life, our brains are rich in Phosphorous and even our DNA contains her, albeit in small amounts.

We mostly profit from her in fertilizers and match ends, though in the 1800s we attempted to use her for medical purposes. In 1874 she was used to attempt cure nervous breakdowns, depression, migraine, epilepsy, stroke, alcoholism, pneumonia, T.B, and cholera: for which she was useless in all cases. Only in the 1930s was she removed from said medicines. In this day and age, however, we are discovering she can be of medical use after all in bone diseases.

Phosphorus, like many mysterious and dark women has enjoyed the attention of poets and writers through the ages. Some had romantic ideas about her luminous splendour and others favoured her more grizzly features and scientific properties. Such romantics included Francis Quarles in 'Emblems of Divine and Moral', A.L. Tennyson in 'In Memorial', Keats 'Lamina' and in A. C. Doyle's 'Hound of the Basker-villes'. To this day she holds the imagination and attention of those in artistic circles enjoying their devotion whether it is through fear or awe, and has evoked much the same adoration in those of a more scientific background.

15 P

Phosphorus

Sulphur 16

Name: derived from the Sanskrit 'sulvere' an ancient name for Sulphur

Subject Notes:

Sulphur is a notorious Element for the following reasons: her rotting egg aroma, and the destruction she instigates. This Element has fiery origins and one can usually smell her presence long before one sees her. In fact if something smells abhorrent or repulsive it is more than likely down to this individual. Examples include a skunk's unique odour, the smell of putrefaction, actual rotting eggs, the emanation from swamps and even halitosis.

Sulphur is often found wandering the streets and alleys selling yellow 'Flowers of Sulphur', which is a crystal like form of Sulphur (orthorhombic crystals). She also begs other Elements for electrons (of which she would like three) and will easily bond or share with others if given the chance. So desperate has this Element become that she will even bond with herself.

Between her odious stench and fiery nature (mentioned in the Bible 15 times, mostly associated with the destruction of Sodom and Gomorrah) she has gathered a rather bad reputation. This has not been helped by her constant use in weaponry, from as far back as Greek fire, right up to gunpowder and beyond in mustard gas. All facts considered this is a rather unfair conclusion. All living things need Sulphur to live. Because of her need for electrons, she is willing to form simple reactions that form amino acids: the building blocks of life.

She can kill bacteria and is extremely good at preserving items, and as a result is added to many of our foods.

Sulphur is not toxic by her self, despite popular belief, but can easily be persuaded by the promise of an electron. For example under the ever corrupting influence of Oxygen she creates Sulphur Dioxide (SO_2) which irritates the nose and lungs and can be the cause of asthma and histamine. It is regretable to think such a good hearted Element could be so easily led astray.

Subject History:

Sulphur may seem to be a scruffy and unkempt unfortunate Element but, as we have discovered behind that dishevelled visage lays a genius responsible for life and the preservation of it, mentioned in the Bible and Homer's Odyssey.

One of the first Elements to be discovered by mankind, almost every country has their own name for her (English Brimstone, Iwo in Japan, Gundhuk in Hindi and the Maori Whanariki, are just a few).

S 16

Sulphur

Selenium 34

Name: From the Greek for Moon.

Subject Notes:

Though Selenium is classed as a Non-Metal she could be put into the Metalloid category, as she possesses schizophrenic tendencies and can exist in two forms: a silvery
metal like substance and a red powder popular in red paint.

As we have seen in this volume of 'The Illustrated Guide to the Elements' many of
the Elements can be integral to human life, and also toxic if administered wrongly,
none more so that Subject 34. This Element is full of contradictions and seems to
pick and choose what she will fight for. If we lack Selenium we manifest symptoms
such as anaemia, high blood pressure, infertility, arthritis and premature aging.
Nonetheless too much will have toxic effects such as loss of hair, a foul body odour
and bad breath to name a few. When dealing with Selenium you walk on a knife
edge. One can't help but feel she enjoys the uncomfortable feeling and uncertainty of
the victims of her attention.

Another seemingly contradictory attribute that Selenium displays is that though
toxic to us in some cases she seems loathe to allow other Elements of a less savoury
nature to poison us. She has earned her self the title as an 'Antagonist' to certain
toxic metals such as Arsenic, Cadmium, Mercury and Thallium, the effects of which
she can counteract.

Subject History:

Other than being vital to our bodies we also use her in paints, plastics, enamels,
photocopiers and solar panels.

Perhaps the most engaging aspect of Selenium's history (and is still a problem to this
day), is the 'Loco Weed' phenomenon. Plants high in Selenium become addictive to
cattle and have an increasingly negative effect. It causes them to stumble, developed
sores, anorexia, fevers and drives them to madness, but because they enjoy the high
they keep on consuming the animal form of meth.

Her name being derived from the Greek for moon is apt since from this word the
word lunatic is derived.

Notes

The Periodic Table of Elements

1	2	3	4	5	6	7	8	9	10	11	12	13	14	15	16	17	18
H 1 Hydrogen																	He 2 Helium
Li 3 Lithium	Be 4 Beryllium											B 5 Boron	C 6 Carbon	N 7 Nitrogen	O 8 Oxygen	F 9 Fluorine	Ne 10 Neon
Na 11 Sodium	Mg 12 Magnesium											Al 13 Aluminium	Si 14 Silicon	P 15 Phosphorus	S 16 Sulphur	Cl 17 Chlorine	Ar 18 Argon
K 19 Potassium	Ca 20 Calcium	Sc 21 Scandium	Ti 22 Titanium	V 23 Vanadium	Cr 24 Chromium	Mn 25 Manganese	Fe 26 Iron	Co 27 Cobalt	Ni 28 Nickel	Cu 29 Copper	Zn 30 Zinc	Ga 31 Gallium	Ge 32 Germanium	As 33 Arsenic	Se 34 Selenium	Br 35 Bromine	Kr 36 Krypton
Rb 37 Rubidium	Sr 38 Strontium	Y 39 Yttrium	Zr 40 Zirconium	Nb 41 Niobium	Mo 42 Molybdenum	43	Ru 44 Ruthenium	Rh 45 Rhodium	Pd 46 Palladium	Ag 47 Silver	Cd 48 Cadmium	In 49 Indium	Sn 50 Tin	Sb 51 Antimony	Te 52 Tellurium	I 53 Iodine	Xe 54 Xenon
Cs 55 Caesium	Ba 56 Barium	Lu 71 Lutetium	Hf 72 Hafnium	Ta 73 Tantalum	W 74 Tungsten	Re 75 Rhenium	Os 76 Osmium	Ir 77 Iridium	Pt 78 Platinum	Au 79 Gold	Hg 80 Mercury	Tl 81 Thallium	Pb 82 Lead	Bi 83 Bismuth	Po 84 Polonium	At 85 Astatine	Rn 86 Radon
~~Fr 87 Francium~~	Ra 88 Radium																

Lanthanoids

La 57 Lanthanum	Ce 58 Cerium	Pr 59 Praseodymium	Nd 60 Neodymium	Pm 61 Promethium	Sm 62 Samarium	Eu 63 Europium	Gd 64 Gadolinium	Tb 65 Terbium	Dy 66 Dysprosium	Ho 67 Holmium	Er 68 Erbium	Tm 69 Thulium	Yb 70 Ytterbium

Actinoids

Ac 89 Actinium	Th 90 Thorium	Pa 91 Protactinium	U 92 Uranium									Pu 94 Plutonium	

The Basics of Elemental Physiology

To understand the Elements one first must understand their biology. Humans have DNA which gives them characteristics that they pass down as family traits. Elements have individual atomic structures.

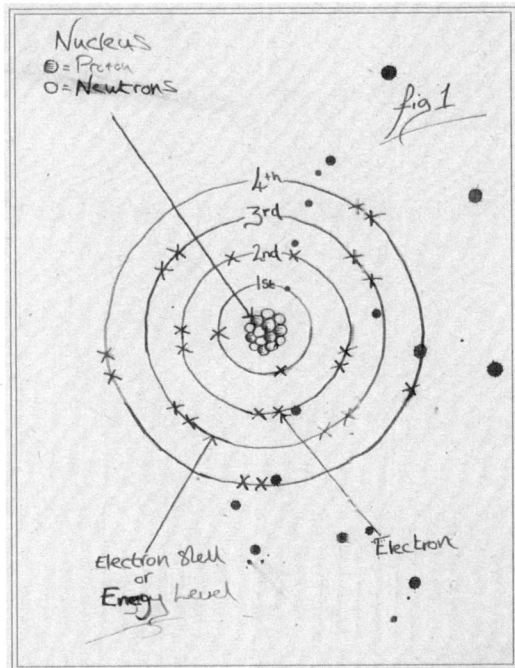

The atomic structure of an Element dictates her size, the state which she lives and how 'electron hungry' she is.

An atom is made of three main parts: Protons, Neutrons and the ever famous Electrons. Protons and Neutrons form the Nucleus (See fig. 1).

Protons are positively charged, and thus attract the negatively charged Electrons. Neutrons are neutral.

Electrons move around the Nucleus in shells and have very little mass. Electrons and Shells are the very base of Chemistry as we know it. The number of Electrons in an Atom is always equal to that of Protons. However, an Element is only happy (i.e. stable) when all of its shells are full. In most atoms this is rare; and this is the cause of reactions.

Electrons can be freely traded between Atoms, and Elements have different ways of getting what they want, gaining losing or sharing electrons. When gaining or losing an electron they become charged one way or the other.

How reactive an Element is, is dependent on how far away the electrons are from the nucleus, and how full or empty their shells already are.

Glossary of Terms

Creep: The tendency of a solid material to slowly move or deform permanently under stress. All ways increases with higher temperatures.

Electronegative: The more electronegative the more powerfully it attracts electrons

Refractory Metals: Metals with a very high resistance to heat and water.

Tensile Strength: The force required to pull to breaking point.

'Aqua Regia': "King of Waters": mixture of Nitric and Hydrochloric Acids.

CNO Cycle: Or Carbon- Nitrogen- Oxygen cycle. It is a catalytic cycle self sustaining, and the dominant source of energy in the stars.

Caesium Clock: A type of atomic clock that uses the frequency of radiation absorbed in changing the spin of electrons in caesium atoms.

Blue Stocking Woman: An intellectual or literary woman. Originally from 'The Blue Stockings Society' which was an informal women's social and educational movement in England in the mid-18th century.

Borate/ Boride: A salt or boric acid/ a compound in which Boron is the most electronegative element.

Pyrophoric: A pyrophoric substance is a substance that will ignite spontaneously in air

Phylum: Family or Group

Helium II: Liquid Helium existing as a superfluid below the transition point of approximately 2.2°K at 1 atmosphere and having extremely low viscosity and extremely high thermal conductivity.

Index